MW01177867

WHEN DINOSAURS RULED THE EARTH

THE TRICERATOPS

Written by Tracy Vonder Brink

Illustrated by Riley Stark

TABLE OF CONTENTS

A Crabtree Seedlings Book

Crabtree Publishing

crabtreebooks.com

School-to-Home Support for Caregivers and Teachers

This book helps children grow by letting them practice reading. Here are a few guiding questions to help the reader with building his or her comprehension skills. Possible answers appear here in red.

Before Reading:

- What do I think this book is about?
    - *I think this book is about dinosaurs.*
    - *I think this book is about the dinosaur called* Triceratops.
- What do I want to learn about this topic?
    - *I want to learn how long* Triceratops *was.*
    - *I want to learn what* Triceratops *ate.*

During Reading:

- I wonder why...
    - *I wonder why paleontologists study fossils.*
    - *I wonder how* Triceratops *used its horns.*
- What have I learned so far?
    - *I have learned that dinosaurs lived before people.*
    - *I have learned that* Triceratops *lived in North America.*

After Reading:

- What details did I learn about this topic?
    - *I have learned that* Triceratops *pulled up plants with its beak.*
    - *I have learned that* Triceratops *had a neck frill.*
- Read the book again and look for the glossary words.
    - *I see the word **paleontologists** on page 4 and the word **herbivore** on page 13. The other glossary words are found on page 22.*

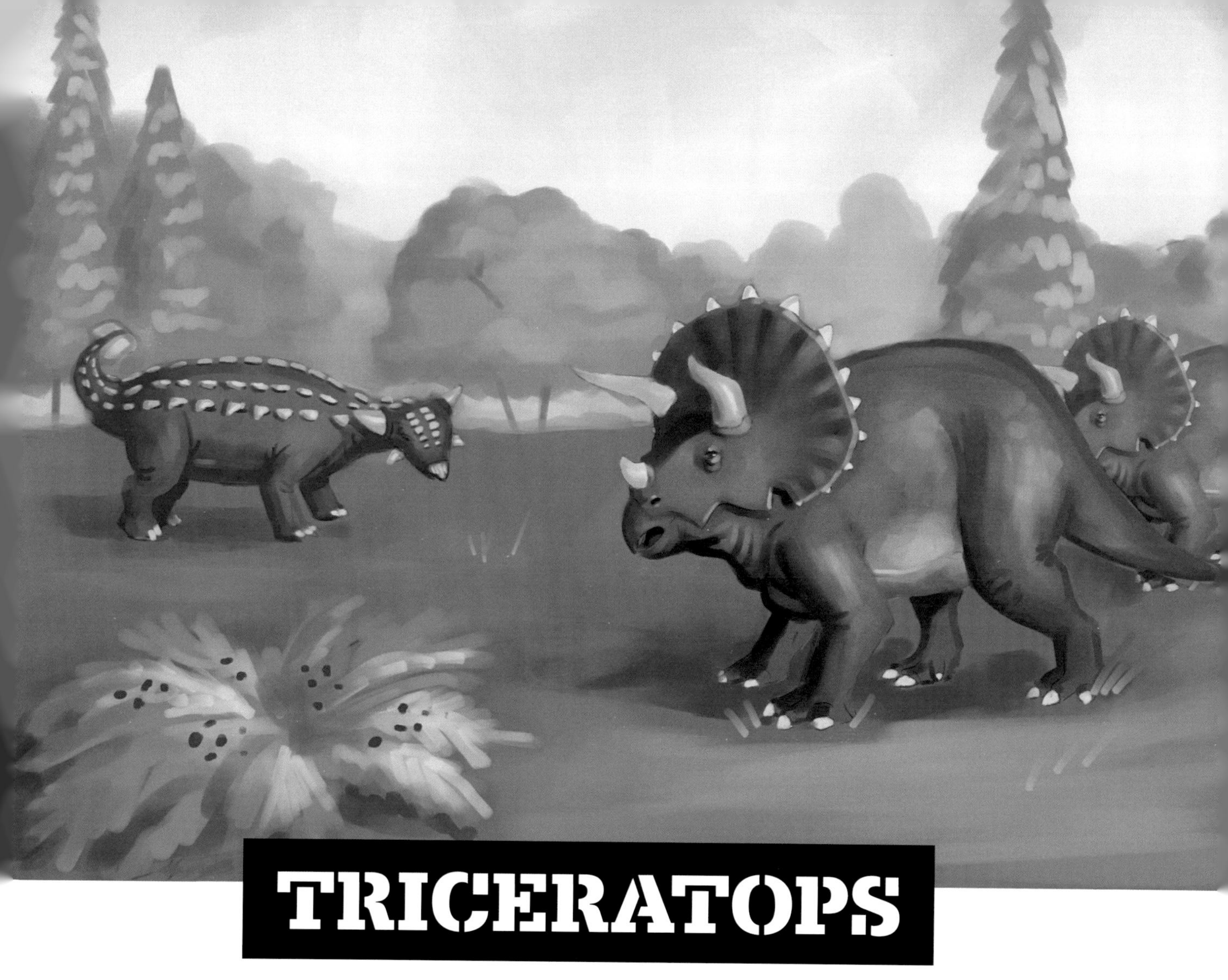

TRICERATOPS

Many dinosaurs once roamed Earth.

They lived long before people.

Some dinosaurs became **fossils** after they died.

Paleontologists study fossils to learn about dinosaurs.

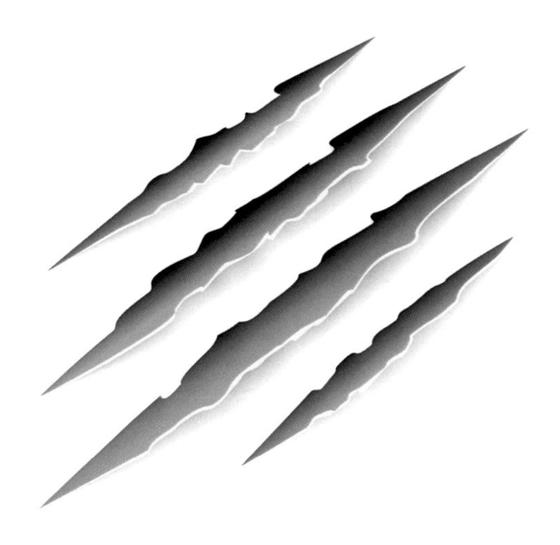

Triceratops was a dinosaur that lived about 68 million years ago.

Triceratops fossils have been found in North America.

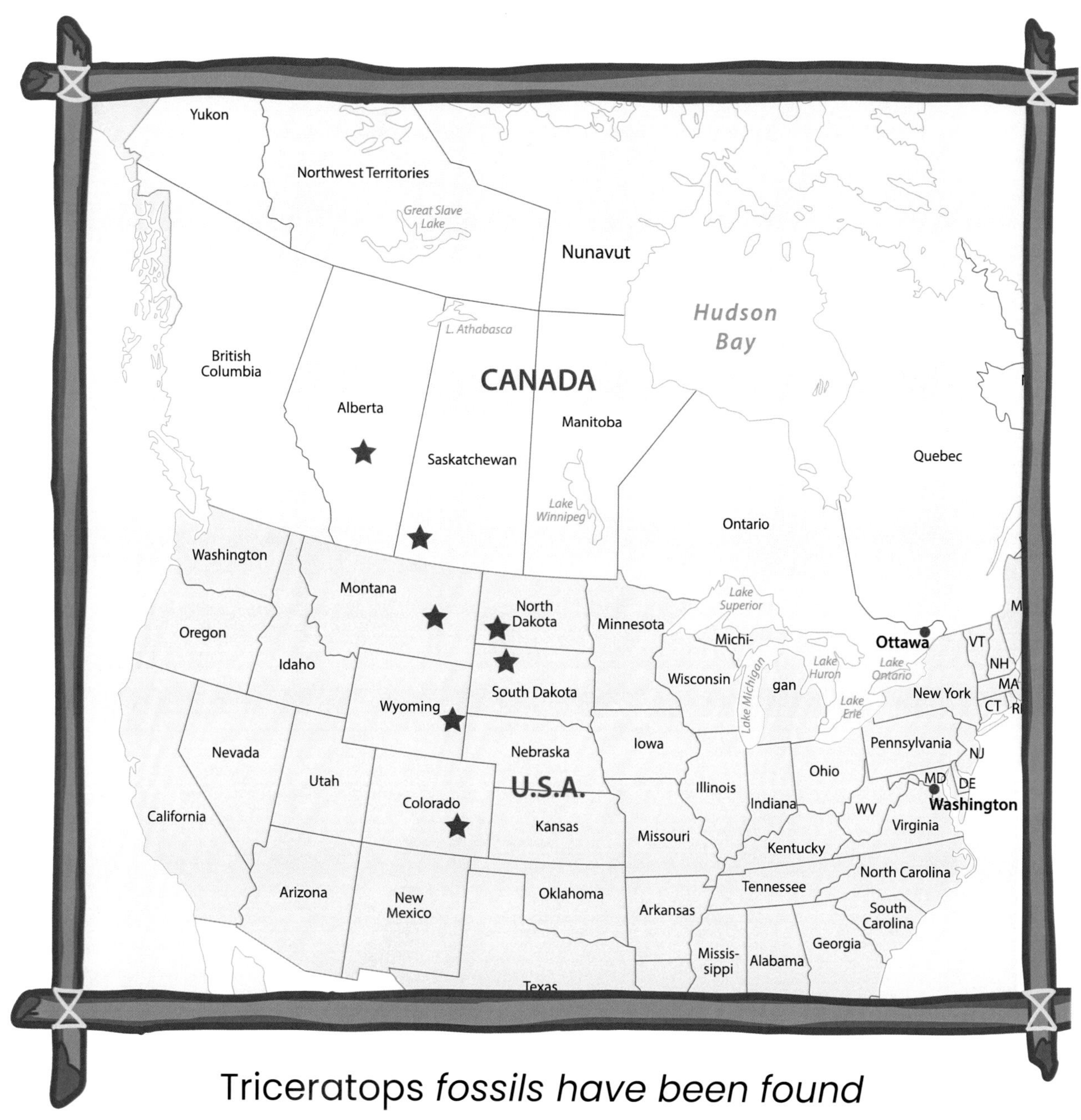

Triceratops *fossils have been found in the United States and Canada.*

Triceratops grew up to 30 feet (9 m) long.

That’s about the size of an African elephant.

Triceratops walked on four legs.

Its back legs were longer than its front legs.

Triceratops was a **herbivore**.

It had a beak, which it used to pull up plants to eat.

Triceratops *ate shrubs, ferns, and other low-lying plants.*

Triceratops had up to 800 teeth.

Its many teeth helped it chew up tough plants.

Triceratops *means "three-horned face."*

Triceratops had two long horns above its eyes and a smaller horn on its **snout**.

It used them to fight off **predators**.

Triceratops had a neck frill that was up to 3 feet (1 m) long.

Did its frill protect its neck?

Triceratops *may also have used its frill to* ***attract*** *a mate.*

So far no Triceratops *fossils have been found grouped together in a herd.*

Did *Triceratops* live in **herds**?

More fossils need to be discovered to find the answers.

Glossary

attract (UH-trakt): To cause interest or attention

fossil (FAWS-uhl): The traces, prints, or remains of plants and animals that lived long ago

herbivore (HER-bi-vor): An animal that only eats plants

herd (HURD): A group of animals that live and feed together

paleontologist (PAY-lee-en-TAW-luh-jist): A scientist who studies fossils to learn about past life on Earth

predator (PREH-duh-ter): An animal that hunts and eats other animals

snout (SNOUT): The long nose of some kinds of animals

Index

About the Author

Tracy Vonder Brink loves to learn about science and nature. She has written many nonfiction books for kids and is a contributing editor for three children's science magazines. Tracy lives in Cincinnati, Ohio, with her husband, two daughters, and two rescue dogs. She wonders what color the neck frills on *Triceratops* were.

Written by: Tracy Vonder Brink
Illustrated by: Riley Stark
Designed by: Rhea Wallace
Series Development: James Earley
Proofreader: Melissa Boyce
Educational Consultant: Marie Lemke M.Ed.

Photographs:
Shutterstock: Evgeny Haritonov: p. 5

Crabtree Publishing

crabtreebooks.com 800-387-7650

Printed in the U.S.A./022024/PP20240115

Published in Canada
Crabtree Publishing
616 Welland Ave.
St. Catharines, Ontario
L2M 5V6

Published in the United States
Crabtree Publishing
347 Fifth Ave
Suite 1402-145
New York, NY 10016

Library and Archives Canada Cataloguing in Publication
Available at Library and Archives Canada

Library of Congress Cataloging-in-Publication Data
Available at the Library of Congress

Hardcover: 978-1-0396-9647-1
Paperback: 978-1-0396-9754-6
Ebook (pdf): 978-1-0396-9968-7
Epub: 978-1-0396-9861-1